GW01218480

FREQUENTLY ASKED QUESTIONS ABOUT
WESTERN SAND DUNES

ROSE HOUK

WESTERN NATIONAL PARKS ASSOCIATION | TUCSON, ARIZONA

SEVERAL MAJOR DUNE fields occur in the western deserts of North America.

Where can I see sand dunes?

Sand dunes are found around the world. Several major dune fields occur in the western deserts of North America. This book focuses on them, primarily the ones in national parks, but much of the geology discussed here is relevant to all dunes.

Where in the western United States can I see sand dunes?

One of the best national parks in which to see dune fields is Death Valley National Park in California. The best-known and most accessible dune field is near Stovepipe Wells; more remote are the Eureka, Panamint, and Ibex dunes. Kelso Dunes are located in Mojave National Preserve, also in eastern California. White Sands National Monument in New Mexico is a single large field of striking white dunes consisting of almost pure gypsum. Great Sand Dunes National Park and Preserve in Colorado features a 30-square-mile active dune field, an adjoining sand sheet, and a cemented-sand deposit called a *sabkha*.

Other significant bodies of sand in North America exist outside national parks. They include the Algodones (Imperial) Dunes in southern California and the Mohawk Dunes across the Colorado River in southern Arizona; Dumont Dunes near Death Valley; Cadiz Dunes not far from Mojave National Preserve; Coral Pink Sand Dunes State Park and Little Sahara Dunes in Utah; Kilpecker Dunes in Wyoming; Sand Mountain in Nevada; and the extensive Gran Desierto in northern Mexico.

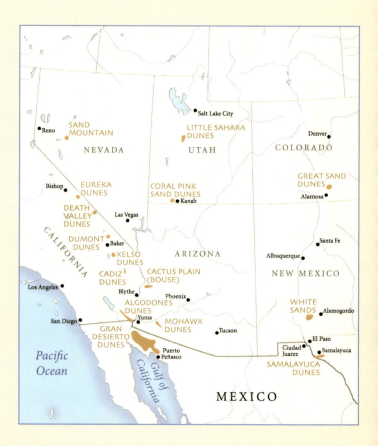

Where are dune fields in other parts of the world?

Along with the Empty Quarter of the Arabian Peninsula, other vast "sand seas" exist in the Sahara, Namib, and Kalahari deserts in Africa; the Taklimakan Desert in China; and in Australia.

EUREKA DUNES

ROUNDED
SAND GRAINS

What does a sand grain look like?

Sand grains commonly are disk, cigar, or sphere shaped. Depending upon the rock they came from and how far they've been moved, the grains can have angular to rounded edges. As wind blows sand around, individual grains ricochet off one another and tend to become pitted and frosted, much like your windshield after it's blasted by a dust storm or sandstorm.

THE SAN ANDREAS MOUNTAINS TO THE WEST OF WHITE SANDS NATIONAL MONUMENT

SAND PARTICLES FROM
GREAT SAND DUNES NATIONAL
PARK AND PRESERVE

What is a sand dune?

A dune is a hill or mound formed by the accumulation of sand. Water can move sand and form underwater or shore dunes, but wind is the primary agent responsible for inland sand dunes.

What is sand?

Geologists define sand not by what it is made of, but by its size. Particles that range between 0.002 inch (0.062 mm) and 0.08 inch (2.00 mm) in diameter are classified as sand. Larger pieces graduate to gravel and cobble-sized silt and clay, smaller ones descend to silt and clay. Quartz is the most common sand material because it is so abundant and is not easily weathered. But sand may also consist of gypsum, volcanic material, or even small bits of seashells.

SAND

WIND

WINDSTORM AT WHITE SANDS

GREAT SAND DUNES

Does the wind always blow in the dunes?

Wind is a fact of life here, and at times it can be an overwhelming force. But the wind does tend to blow a little harder in the spring months—March, April, and May—with gusts often exceeding 30 miles an hour in the afternoons. With almost no plant life to break the wind, it does indeed seem windier on dunes. If you are out in the dunes, the best place to escape the wind is in the quiet air on the lee face (protected side) of a dune.

How does the wind move sand?

Wind can move sand in several ways. In a motion called "creep," grains hit other grains, nudging them slightly out of place, and the heavier the grain, the closer it stays to the ground. The majority of sand grains, though, bounce or jump along in short hops, a movement termed saltation. In strong winds, saltating grains will form a flurry of sand about a foot or so above the surface of a dune. The smallest grains are carried high into the air in suspension, often mixing with dust.

6

STRONG
enough, it can pluck grains of sand from the ground and transport them for miles.

What's needed to make a dune field?

The key ingredients are a source of sand, wind sufficient to pick up the sand and move it, and changes in wind direction and velocity. Topographic features, weather patterns, plants, and other features can affect the deposition of sand. When the wind is strong enough, it can pluck grains of sand from the ground and transport them for miles. When sand encounters an obstacle, say a rock, the grains fall to the ground and begin to accumulate around the obstruction. Over time the obstacle is buried by a building dune, which itself obstructs the wind. Where a barrier exists downwind from the source of sand, as occurs with the Sangre de Cristo Mountains near Great Sand Dunes, a field of dunes can form.

What is the source of the sand?

Sand grains are tiny pieces of rock—granite, sandstone, or basalt, for example. The parent rock is broken down by mechanical and chemical weathering into boulders and even smaller pieces. Flowing water deposits this material at the foot of mountains and in valley lakes. Wind steps in to pick up the loose particles and sorts them by size: Smaller grains stay suspended in the air longer and are carried farthest from their source. Sand also ends up in rivers and streams. Some is left behind on shorelines or in dry streambeds, and the wind picks it up later. In other cases, sand comes from the deposits in dry lakes, or playas.

THE KELSO DUNES ARE FED SAND FROM THE MOJAVE RIVER SINK AND SODA DRY LAKE (AERIAL)

THE SANGRE DE CRISTO MOUNTAINS PROVIDE A BARRIER AT GREAT SAND DUNES (AERIAL)

SELENITE CRYSTALS NEAR LAKE LUCERO, WHITE SANDS

The immediate source of sand is different for each dune field. For the dunes near Stovepipe Wells in Death Valley, the sand likely comes from the foot of the Cottonwood Mountains and a dry lake northwest of the dune field. A closed basin northwest of Great Sand Dunes is the closest sand supply for the dunes in this national park. The Kelso Dunes, cradled between the Granite and Providence mountains, are fed sand from the Mojave River Sink and Soda Dry Lake to the west. The nearly pure gypsum sand at White Sands comes from eroding selenite crystals in Lake Lucero, a unique large playa on the west side of the park, and from sediments in the nearby mountains.

To discover ultimate sources of sand, geologists look at the chemical and mineralogical "fingerprints" of the grains. At Great Sand Dunes, for example, volcanic sand fragments point to an origin in the volcanic rocks of the San Juan Mountains. Grain texture—size, shape, and degree of roughness—can also reveal whether the grains have been carried relatively short or long distances. By geologic standards sand dunes are young, dynamic features, moving like the wind. But the rock from which the sand was derived could be millions of years to a billion years old.

WIND

Slipface

Rolling grains land on slipface

Unstable accumulation builds up

Accumulation cascades down to base, advancing the dune

GREAT SAND DUNES (LEFT)
WHITE SANDS (CENTER)
DEATH VALLEY, AERIAL (RIGHT)

How fast do sand dunes move?

Some dunes move slowly, mere inches a year, while others zip along at as much as 50 to 80 feet a year. Other dunes show little net movement and instead grow vertically.

CRESCENT DUNE

or *barchan*, the Arabic word for a ram's horn

Are all sand dunes alike?

No. Sand dunes assume different shapes depending upon sand supply; plant life; and wind direction, consistency, and velocity. Several major types of dunes have been defined, and different types may exist in one field. One is the crescent dune, or *barchan*, the Arabic word for a ram's horn, which aptly describes the shape. Crescent dunes result from moderate winds that come predominantly from one direction. A *parabolic* dune is the same shape as a *barchan* only in reverse, with the arms extending upwind instead of downwind. *Transverse* dunes are linear ridges arranged at right angles to the prevailing winds. *Star dunes*, as the name implies, have arms radiating out like stars. They are created when wind blows from varied directions at different times of the year. *Dome-shaped* dunes are small, fast-moving circular mounds. They are fairly rare but can be seen at White Sands near Lake Lucero, charging northeast and merging with other dunes in the heart of the park.

Because sand is derived from different rock types, it comes in different colors as well, from palomino to pink, tan, coral, white, or black. The colors change with shifting light and shadows throughout the day. Mirages, those wavy visual effects on the horizon caused by differing air densities, make the dunes appear to undulate and affect the apparent color.

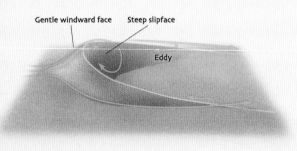

Barchan dune

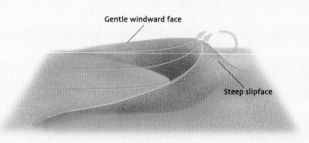

Parabolic dune

BEAUTIFUL RIPPLE  marks often etch the sand surface.

Do dunes continue to build and migrate? What can stop them?

Dunes will continue to build and migrate as long as a source of adequate sand is available, there is wind to move them, and nothing gets in their way. A mountain range, reversing wind patterns, or flowing streams, as at Great Sand Dunes, for example, can serve as barriers to migrating dunes. Sand in a dune field is often recycled, migrating until it reaches a barrier, then cycling back down a streambed or a lakebed, and finally becoming windborne again once the water leaves it behind. This process may take some time, often on the scale of decades or centuries, and the cycling rate depends on changing wind and weather patterns.

WIND RIPPLE, WHITE SANDS

How do ripples form on dunes?

Beautiful ripple marks often etch the sand surface. They are actually miniature dunes riding on the back of a larger dune. Ripples form as local winds blow saltating grains along the surface; the sand grains form crests and troughs perpendicular to the wind direction. Ripples form instantaneously and change spontaneously. You can smooth them away with your hand and watch them reform in minutes if the wind is blowing.

GRAN DESIERTO, MEXICO

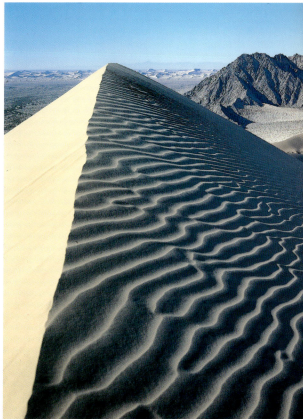

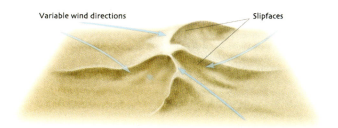

Star dune

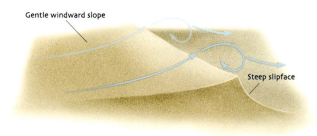

Transverse dune

What is the angle of repose of sand?

Dry sand grains gather on the lee side of a dune until the slope reaches 30 to 34 degrees, called the angle of repose. Pile sand any higher, and gravity is going to take over. At the angle of repose, either grains of sand start to slump down a dune's lee side or entire tongues of sand slump or avalanche down. If you try to climb *up* the lee face, you'll soon appreciate how easily the sand slides and how difficult it is to gain a foothold.

DUNE

IBEX DUNES, DEATH VALLEY

CROSSBEDDED SANDSTONE

What is the largest dune field in North America?

The Gran Desierto in northern Mexico, at 1,700 square miles, is the largest dune field in North America. In comparison, the largest sand sea in the world is the Empty Quarter of the Arabian Peninsula, some 225,000 square miles.

What is the tallest dune in the United States?

The Star Dune in Great Sand Dunes National Park and Preserve rises to about 750 feet above the surrounding terrain. Elsewhere in the West, some active dunes are small mountains of sand approaching 600 feet high. Because active dunes are always moving and changing, "tallest," "widest," or any other such superlatives are subject to challenge when applied to dunes.

What is the anatomy of a sand dune?

In general, a single dune has a gentle, curved slope on the windward side. Sand grains blow up this slope until they reach the crest, or brink, where they rapidly drop onto the lee slope, or slipface, of the dune. On the windward side the sand may be firm, but on the steeper slipface it is soft and loose, given to avalanching. If the wind direction shifts periodically, it produces elegant crossbedded layers in the sand, seen if a trench is cut into a dune. These crossbeds are preserved when sand turns to sandstone, and they provide evidence of ancient wind patterns.

STABLE DUNES, WHITE SANDS

What is a stable dune?

A stable dune hardly moves at all. Stable dunes are held in place by the roots of woody plants, they often have a crust that decreases water infiltration, and they may occupy the margins of an active dune field.

ACTIVE DUNE, DEATH VALLEY

What is an active dune?

An active dune is moving or building vertically. It is essentially any dune whose sand is available to be blown by the wind because nothing is cementing or sheltering it. As sand is eroded from the windward slope of a dune and deposited on the lee face, the dune keeps moving downwind.

STAR DUNE, GREAT SAND DUNES, AERIAL

WIND

Former position   Erosion   Future position   Deposition

SAND IN A DUNE FIELD is often recycled, migrating until it reaches a barrier, then cycling back down a streambed or a lakebed, and finally becoming wind-borne again once the water leaves it behind.

SCIENCE

GREAT SAND DUNES

It is an iron-rich mineral called magnetite. If you run a magnet across the sands, it will pick up the fine magnetite particles.

CEMENTED SAND,
DEATH VALLEY

What tools do scientists use to study dunes?

Anemometers, sand traps, erosion stakes, surveying instruments, satellite photographs, wind tunnels, microscopes, maps, and seismographs all help in research on eolian processes—that is, any process driven by the wind. Anemometers measure wind speed, and sand traps and erosion stakes reveal sand movement. Wind tunnels and microscopes are used in the laboratory to study wind behavior and sand grain mineralogy. Scientists are still working out some puzzles about dunes: Why are they oriented in certain directions? What determines their size, spacing, and patterns? What is the source of the sand? How old are the dunes? They are even looking to other planets, such as Mars and Venus, that exhibit eolian processes.

A pioneer in sand studies was British Army Brigadier Ralph Bagnold, who spent a decade in the deserts of Egypt in the 1920s. His book, *The Physics of Blown Sand and Desert Dunes*, is considered a classic. Edwin McKee, a scientist with the U.S. Geological Survey, added to our knowledge with his research at White Sands, Great Sand Dunes, and other dune fields in the mid-twentieth century.

IN THE KELSO DUNES, for instance, one layer is estimated to be 25,000 years old.

How old are dunes? How are they dated?

In general, some dune fields in the West apparently started to accumulate within the last 30,000 years. During the last glacial period, when big Ice Age lakes were filled and rivers were flowing year-round, the sand supply was not as abundant. As the climate warmed, however, the rivers got smaller or became intermittent, and lakes evaporated. Sediments dried out and were more readily transported by wind. The full answer is more complicated, however, because in some dune fields sand was deposited at different times. In the Kelso Dunes, for instance, one layer is estimated to be 25,000 years old. Younger layers were deposited on top of this layer. At Great Sand Dunes, sand in the uppermost layers is 300 to 600 years old, while sand buried hundreds of feet down probably goes back more than 20,000 years. These figures have been obtained largely by an experimental method called optical luminescence dating, which measures how much time has passed since sand grains were last exposed to sunlight.

SNOW AT THE
PAINTED DESERT

Do sand dunes really "sing"?

You could say dunes "sing," but what has been reported more often are squeaking and booming sounds, likened to a cello, squeaky tennis shoes, or barking dogs. Acoustic dunes produce sounds when triggered by human movement such as stomping on the surface or digging a trench in the sand. The sand itself often vibrates with the sounds. The source of the sound may be a combination of compression and slippage of the sand. The phenomenon is aided by a silica coating on the sand grains that helps them stick together and resonate when they're moved. Kelso Dunes in Mojave National Preserve and Eureka Dunes at Death Valley are famous for their reverberations. But under the right conditions, especially after strong spring winds when large slipfaces form, any dune can express itself in sound.

KELSO DUNES,
MOJAVE NATIONAL PRESERVE

PLANT LIFE

BORREGO LOCOWEED, KELSO DUNES

Can plants live on the dunes? How do they survive?

Some plants can live on dunes, but only those with special adaptations are able to surmount the swift burial, drastic excavation, and nutrient-poor soil that occurs on active dunes. Among those adaptations are rapid growth and gigantism, along with a high rate of photosynthesis. Dune buckwheat, for example, speedily elongates its stem so leaves and flowers can stay ahead of burial; stems of dune grass on the Eureka Dunes can lengthen more than 2½ inches in a single day; fourwing saltbush and creosote bush on some dunes can grow to twice their normal three- to four-foot height. Plants also sprout roots along their stems. The adventitious roots of some soaptree yuccas at White Sands approach 30 feet long!

On active dunes, plants also trap sand and build hummocks around their roots, allowing them to survive the strongest erosive winds. Again at White Sands, the gypsum forms giant hummocks beneath the stems of shrubs. The pedestals remain as the dunes continue to move because the shade of the plant and the adventitious roots keep that part of the dune from drying up and crumbling. Underground, various dune plants make up for nitrogen deficits by having special bacteria on their roots that fix nitrogen. Members of the pea family and some dune grasses possess this ability. To withstand the bright, hot conditions of dune fields, leaves are often covered with silvery-gray hairs that retard evaporation. Other plants are adapted to deal with the salt- and gypsum-rich environments present on some dunes.

PEDESTAL PROTECTED BY PLANT AT WHITE SANDS

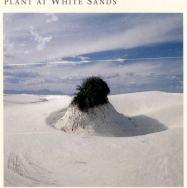

EVENING PRIMROSE AND SAND VERBENA

INTERDUNE CRUST

What are the low places between individual dunes?

Low spots between dunes are called interdunes, troughs, or blowouts. The ones at White Sands are especially well developed because the water table is close to the surface. At that park, moist interdunes support cyanobacteria, or blue-green algae, imparting a greenish tinge to the interdune area. The algae associate with fungi, forming a protective lichen crust.

An interdune's flat, sometimes damp, surface supports a procession of plants, including grasses, evening primrose, sand verbena, and shrubs. Nitrogen "fixed" by cyanobacteria helps these plants grow. Mites and other small soil organisms in the plant litter aid decomposition in this interesting ecosystem. At times standing water collects in the interdunes, attracting shorebirds. Advancing active dunes eventually envelop some interdunes, their movements forming new interdunes elsewhere.

Are sand dunes moist or dry?

Although they look exceedingly dry, many dunes are actually moist below the surface because rain drains down into the sand instead of running off. In addition, the upper layers of dry sand help insulate the subsurface sand and keep it cool and moist. In some dunes a permanently moist layer is found only a foot or two down. At Great Sand Dunes, water plays a complex role as it moves through various parts of the ecosystem. A large regional aquifer of underground water exists beneath the dune field, affecting wetlands and streams within the dune field.

PONDEROSA PINE,
GREAT SAND DUNES

GROWING
CALIFORNIA
CROTON, MOJAVE
NATIONAL PRESERVE

SOAPTREE YUCCA, WHITE SANDS

What are the dead-looking branches that stick out of the dunes?

Although not abundant, a few trees do manage to continue to grow among advancing dunes—honey mesquite at Death Valley, ponderosa pines at Great Sand Dunes, and Rio Grande cottonwoods at White Sands, for instance. These trees suffer when accumulating sand chokes off the oxygen supply to their roots by covering their branches and leaves. Dried branches or dead trunks are all that remain after the trees are buried by a dune. At Great Sand Dunes a "ghost forest" of ponderosa trunks still stands, though the trees themselves are dead.

THE ADVENTITIOUS ROOTS of some soaptree yuccas at White Sands approach 30 feet long!

SHORT-HORNED LIZARD

ANIMAL LIFE

COYOTE

What animals live in the dunes? What signs do we see of them? How have they adapted?

Insects, reptiles, mammals, and birds all live in sand dunes, though they're not often visible. On occasion, you may see the animal itself—perhaps a blue-black raven or a tawny kit fox if you're very lucky. In many dune fields with searing daytime temperatures, animals tend to stay underground where it's cool and moist.

Mostly you will see signs of animals in the form of tracks in the sand or their burrows at the base of shrubs. Beetles and crickets, some of them nearly transparent, leave trails like feathered stitches on a crazy quilt. Lizards are not uncommon but are well concealed by adaptive coloration—species such as the Cowles prairie lizard and bleached earless lizard at White Sands are pale in tone. The fringe-toed lizard is perfectly suited to dune living, with enlarged scales on the toes that allow it to "swim" in the sand. Other lizards and some snakes have shovel-shaped heads for better digging in loose sand. In the hot deserts, the sidewinder rattlesnake inscribes telltale curves as it loops across dunes in a unique mode of locomotion. When not out at night hunting, the sidewinder stays completely concealed under the sand.

Kangaroo rats are about the only mammals that actually live *on* the dunes. They're out at night, bounding along the ground in search of seeds, which they metabolize to supply all the water they need. When they're really moving, kangaroo rats leave tracks only of their hind feet; if they go down on all fours, their tail adds a long slender trace. Kit foxes have hairs between their toes to help negotiate shifting sand. You might also spy the heart-shaped tracks of deer and elk or the doglike imprint of a coyote, visitors to the dunes.

SAND FOOD, A PARASITIC PLANT

TIGER BEETLE, GREAT SAND DUNES

Why are some plants and animal species called endemic?

Plants and animals have developed many adaptations to the unique environment offered by sand dunes. But some species—called endemics—are so specialized that they are found only in one particular dune field. These species have had enough time to evolve in an isolated environment with distinct conditions. The Eureka Dunes in Death Valley, for instance, contain three endemic plants—dune grass, dune locoweed, and dune evening primrose. In the Kelso Dunes, Borrego locoweed is an endemic. In the gypsum dunes of White Sands, plants like gyp nama actually must have gypsum to survive, while other plants are able to tolerate the unique conditions in this dune field. Welsh's milkweed is another endemic, growing only on the Coral Pink Dunes in southern Utah. In the animal world, Great Sand Dunes is home to at least seven endemic insect species, among them the Great Sand Dunes tiger beetle and the circus beetle.

Have people lived in the dunes? Was there anything to eat?

In the past, Native people mostly visited rather than lived on the dunes, coming to harvest edible seeds of ricegrass and other plants. When they stopped to camp, it was often at springs or lakes at the edges of dunes, where they built fires and ground seeds using stone tools. At Great Sand Dunes, Ute and Jicarilla Apache stripped bark off ponderosa trees growing near the dune field, using the bark for shelter, food, and medicine. Some Native groups regard the sands as sacred. At White Sands, the Apache gathered salt from the playas. To the southwest, on the U.S.-Mexico border, the *Hia C'ed O'odham* harvested a parasitic plant called sand food in the Gran Desierto and Algodones dunes. It has a long, fleshy stem and appears like a saucer-sized mushroom on the dune surface in the spring. The O'odham dug and roasted this prized "dune root," said to taste like sweet potato.

HIKER, GREAT SAND DUNES

PINACATE BEETLE

Did people travel through or across dunes?

If they could not find a way to avoid them, people did travel across dune fields, usually with great difficulty. Any road through an active dune field is inevitably engulfed by sand—at White Sands a sandplow, much like a snowplow, is put into service to clear the Dunes Drive. In California's Algodones Dunes an old road made of wooden planks was simply picked up and relocated when sand covered it. Some pieces of it might still be seen just west of Yuma. In the mid-1800s, military surveyors imported camels to the United States, hoping to aid their passage across the sandy terrain. The experiment was short-lived because, among other things, the troops didn't take to the beasts of burden.

GYP NAMA, WHITE SANDS

IN THE GYPSUM

dunes of White Sands, plants like gyp nama actually must have gypsum to survive, while other plants are able to tolerate the unique conditions in this dune field.

PLACES

When is the best time to visit?

Nearly any time is good, but the dunes are especially beautiful to behold at sunrise and sunset, or even by moonlight. Evenings, nighttimes, or early mornings are the times when wildlife might be about. In summer these are also the best times to avoid the searing temperatures.

What do off-road vehicles do to dunes?

Off-road-vehicle use is not allowed on dunes in national parks or preserves. The vehicles can harm plants and animals—literally crushing plants and disturbing or injuring animals. Erosion and runoff are also hastened, damage that requires a long, long time to heal. It is not speed and noise, but silence and a pure beauty that dunes offer us.

Is it safe to visit dunes?

The dunes are irresistible, magical places to visit. They do require some effort to climb, but walking is a bit easier along the crest where the sand is firmer. In summer in the deserts, surface temperatures on the dunes can reach 140 degrees Fahrenheit, and prolonged exposure to such temperatures can kill. Wear shoes. Carry plenty of water. Keep your bearings: Wind can obliterate footprints in short order, and clouds can obscure landmarks. The good news is if you should fall, it's a soft landing.

Do sand dunes contain any valuable minerals?

Quartz sand is the raw material for glass, and gypsum has been mined for plaster and cement. Fine gold deposits interested a prospector at Great Sand Dunes in the early 1900s, but they never proved productive. At Kelso Dunes, people staked claims to the iron-rich magnetite in the sand, but nothing ever came of that either.

Oil and water are often found in sandstones. The sandstone is formed when sand grains are cemented, and layers build up either on land or in the sea. This process can happen quickly on the surface, as in a playa or *sabkha*. Other times it takes millions of years of burial under layers of sediments. The degree of cementation is important because it affects the amount of pore space and permeability where water and petroleum are stored.

EVENINGS, NIGHTTIMES, OR EARLY MORNINGS are the times when wildlife might be about.